On the Go

Subtract.

How many are left?

$6 - 2 = 4$

How many are left?

$9 - 3 =$ ____

How many are left?

$10 - 8 =$ ____

Use a sticker to answer.

What has no hands but can wave all day?

Animal Action

Subtract.

How many are left? ______

$7 - 5 = 2$

How many are left? ______

$9 - 4 =$ ______

How many are left? ______

$10 - 3 =$ ______

How many are left? ______

$8 - 8 =$ ______

Find stickers for the missing numbers.

10, 9, 8, ▽, 6, 5, ☐, 3, 2, 1

Skill: To practice subtraction facts through 10.

page 1

page 2

page 5

page 6

page 3

page 10

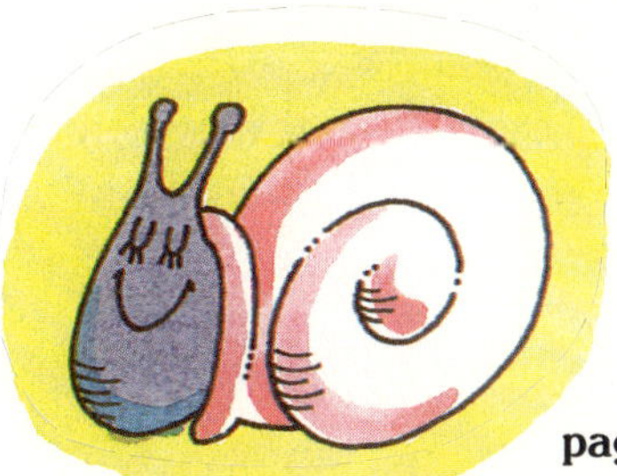

page 7

page 8

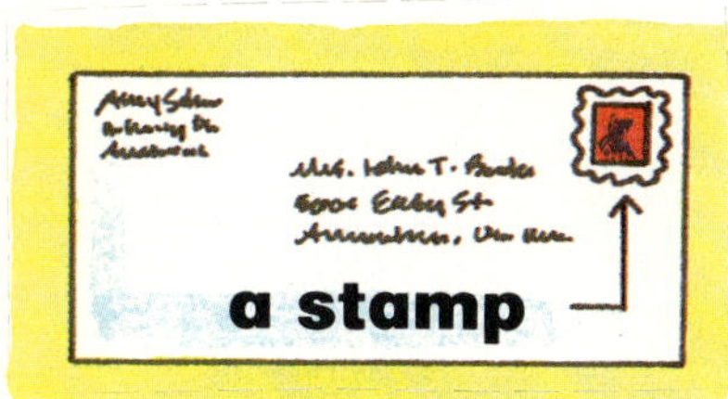

page 9

page 4

page 11

page 14

page 19

page 12

page 13

More stickers at back of book!

More Animals

X out.
Subtract. Write how many are left.

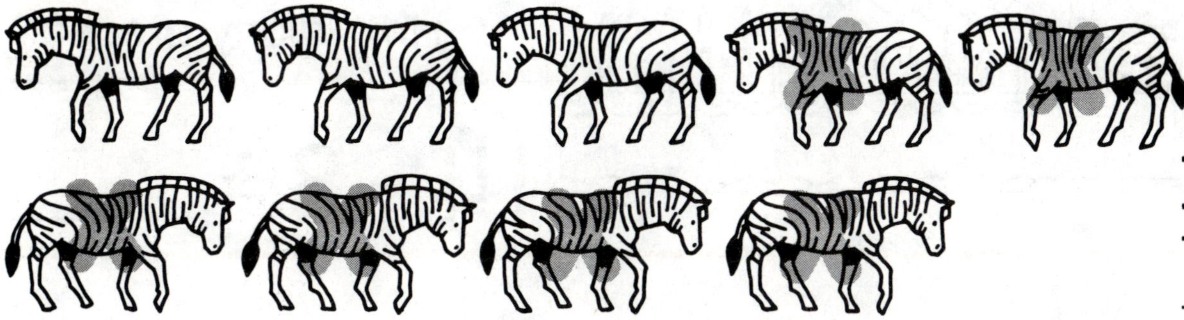

$$\begin{array}{r} 9 \\ -\ 6 \\ \hline \end{array}$$

How many are left? 3

$$\begin{array}{r} 10 \\ -\ 6 \\ \hline \end{array}$$

$$\begin{array}{r} 7 \\ -\ 2 \\ \hline \end{array}$$

$$\begin{array}{r} 8 \\ -\ 5 \\ \hline \end{array}$$

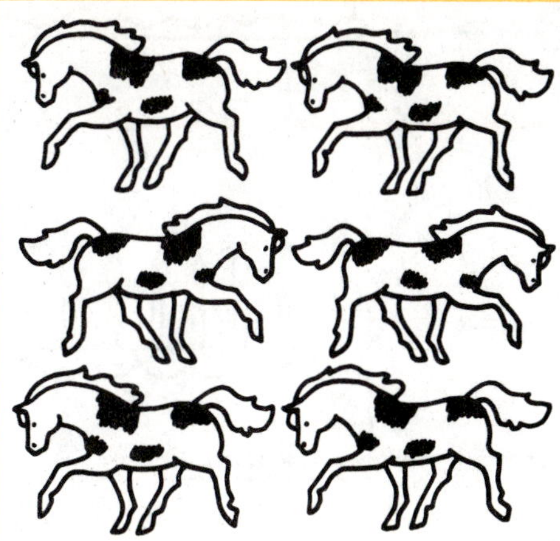

$$\begin{array}{r} 6 \\ -\ 5 \\ \hline \end{array}$$

Use a sticker to answer.

What animal is easy to spot?

Party Time

Write how many are left.

How many are left?

$14 - 8 =$ 6

$13 - 6 =$ ____

$12 - 7 =$ ____

$9 - 0 =$ ____

Use a sticker to complete.

I scream, you scream.
We all scream for . . .

Skill: To subtract numbers 0–9 from numbers through 14.

Don't Bug Me!

Write how many are left.

How many are left?

$14 - 5 = 9$

$13 - 9 = $ ____

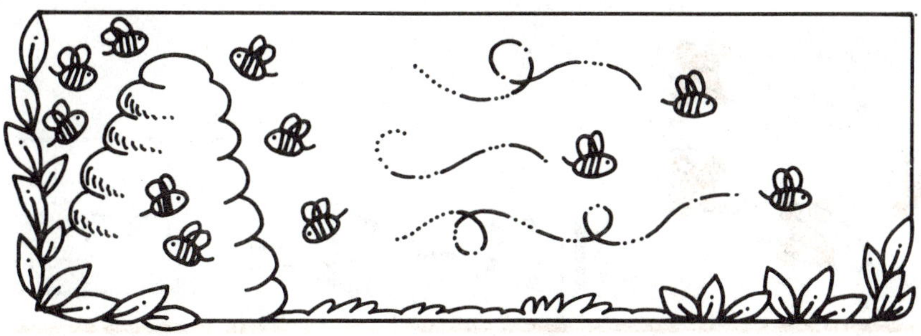

$11 - 3 = $ ____

$12 - 5 = $ ____

Use a sticker to answer.

What has 4 wheels and flies?

Nutty Cross Out

Subtract.
Write how many are left.

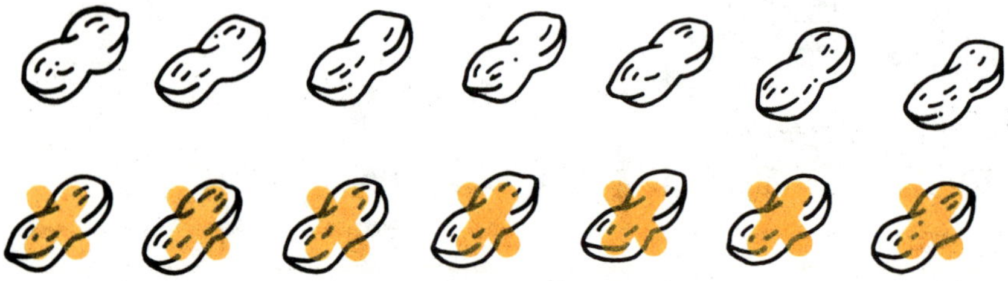

How many are left?

$14 - 7 = $ 7

$11 - 2 = $ ____

$13 - 8 = $ ____

$12 - 9 = $ ____

Use a sticker to answer.

A  **got 10**  **.**

The  **ate 7**  **.**

How many  **are left?**

Skill: To subtract numbers 0–9 from numbers through 14.

Home Sweet Home

X out.
Subtract. Write how many are left.

How many are left?

$14 - 9 =$ 5

$10 - 7 =$ ____

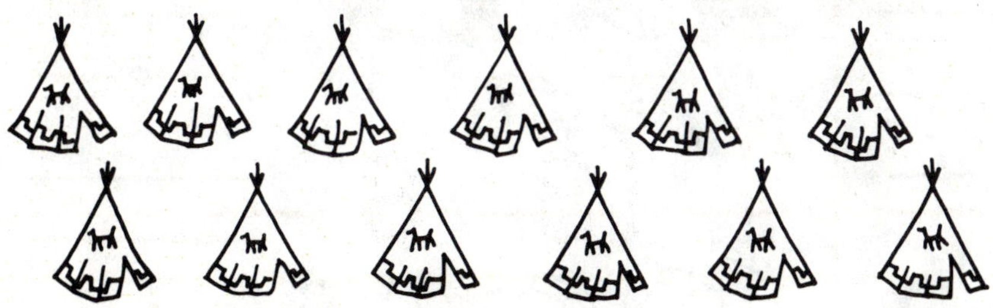

$12 - 6 =$ ____

$13 - 5 =$ ____

Use a sticker to answer.

What animal never leaves home?

Skill: To subtract numbers 0–9 from numbers through 14.

Down We Go

Subtract.
Write how many are left.

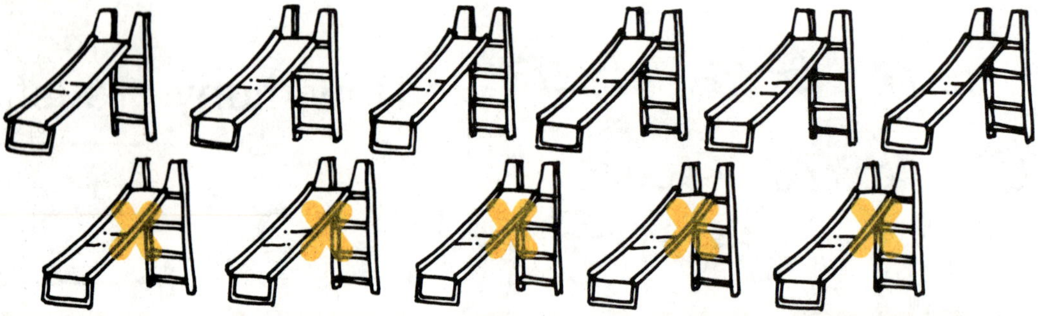

$$\begin{array}{r} 11 \\ -\ 5 \\ \hline 6 \end{array}$$

How many are left? 6

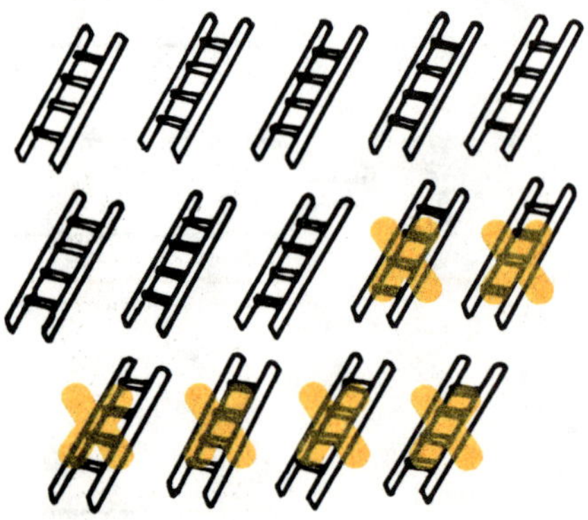

$$\begin{array}{r} 14 \\ -\ 6 \\ \hline \end{array}$$

$$\begin{array}{r} 13 \\ -\ 4 \\ \hline \end{array}$$

$$\begin{array}{r} 12 \\ -\ 8 \\ \hline \end{array}$$

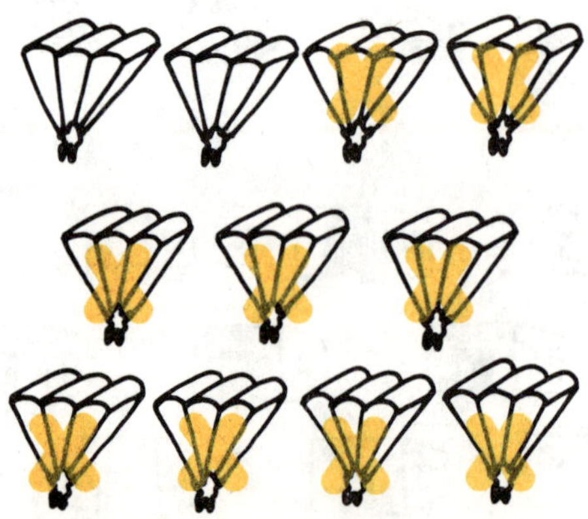

$$\begin{array}{r} 11 \\ -\ 9 \\ \hline \end{array}$$

Find stickers for the missing numbers.

14, 12, ☐, 8, ◯, 4, 2

Skill: To subtract numbers 0–9 from numbers through 14 in vertical form

Have a Nice Trip

X out.
Subtract. Write how many are left.

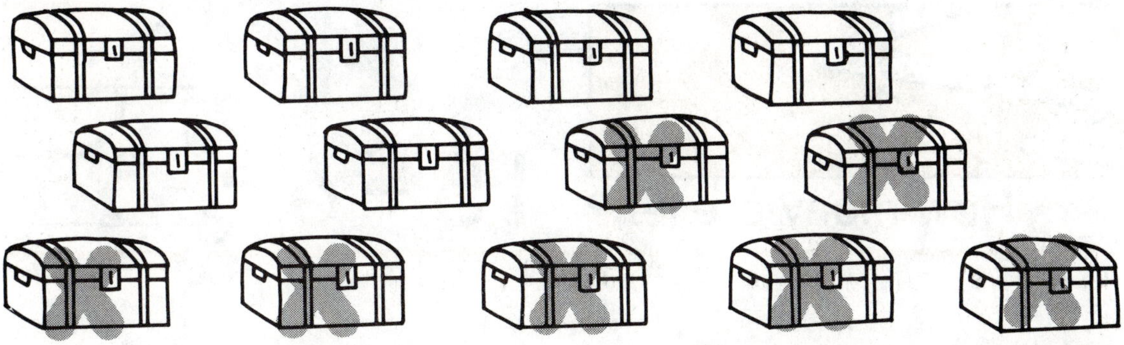

How many are left?

$$\begin{array}{r} 13 \\ -\ 7 \\ \hline 6 \end{array}$$

$$\begin{array}{r} 12 \\ -\ 3 \\ \hline \end{array}$$

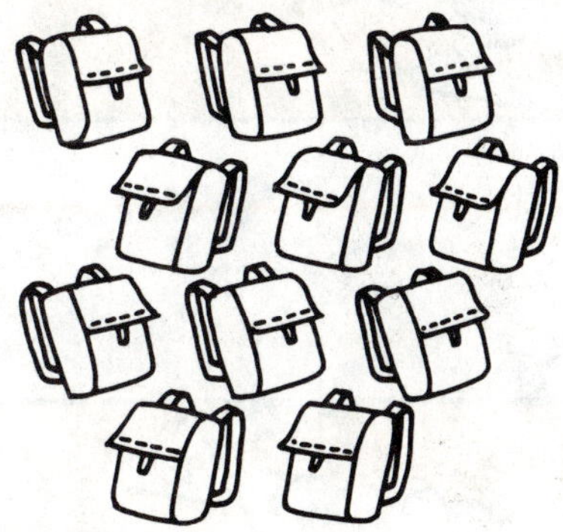

$$\begin{array}{r} 11 \\ -\ 6 \\ \hline \end{array}$$

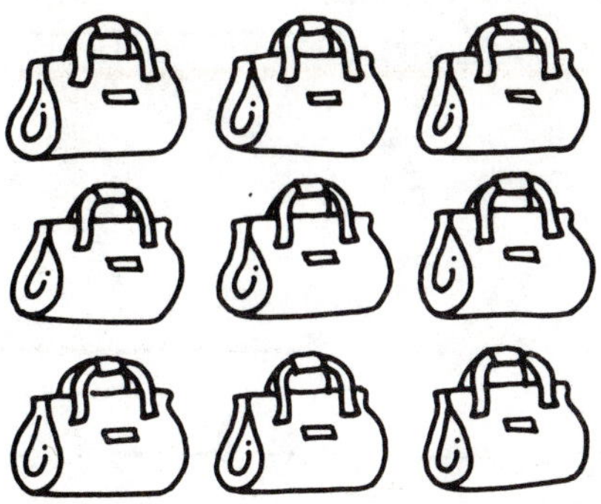

$$\begin{array}{r} 9 \\ -\ 5 \\ \hline \end{array}$$

$$\begin{array}{r} 14 \\ -\ 8 \\ \hline \end{array}$$

Use a sticker to answer.

What can travel all over but never leaves the corner?

Rolling Along

Write how many are left.

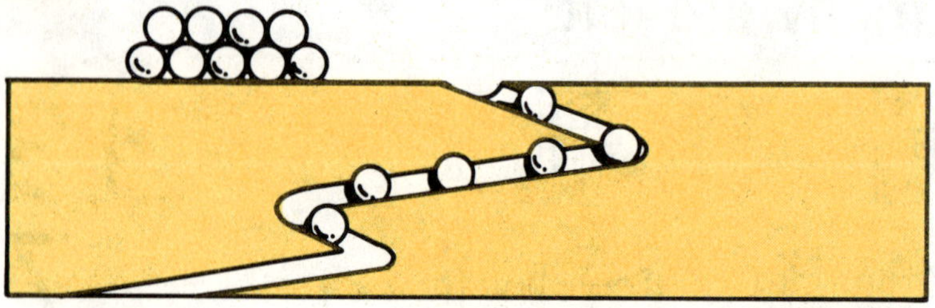

How many are left?

$15 - 6 = 9$

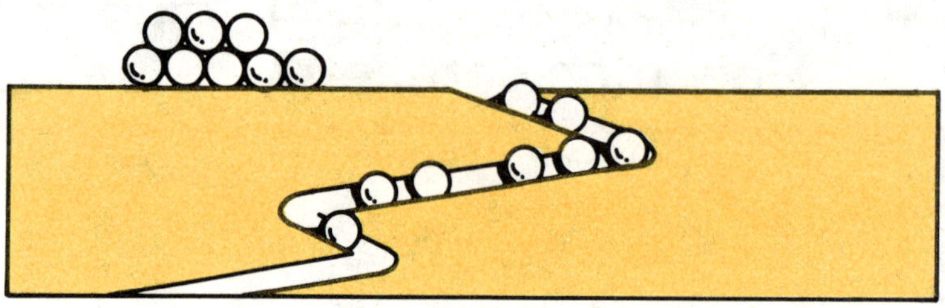

$16 - 8 =$ ____

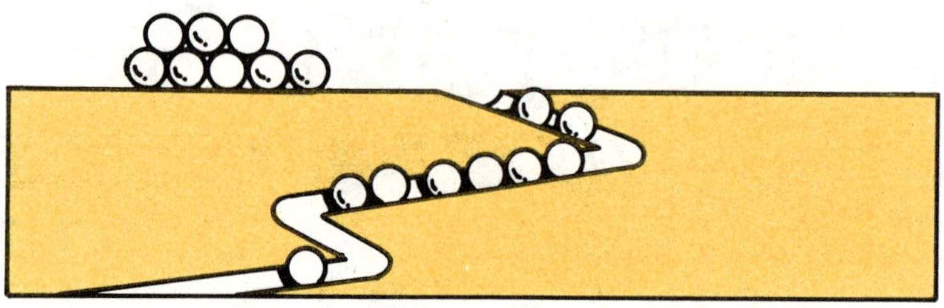

$17 - 9 =$ ____

$14 - 7 =$ ____

Find a sticker that shows how many are left.

Oops!

Skill: To subtract numbers 0–9 from numbers through 18.

Fishy Facts

Write how many are left.

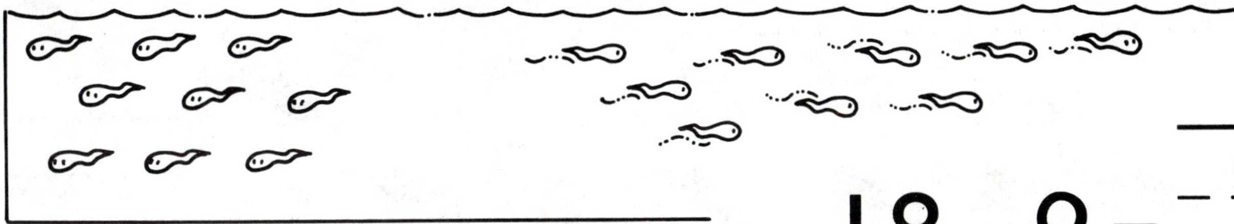

How many are left? $18 - 9 = 9$

$17 - 8 =$ ____

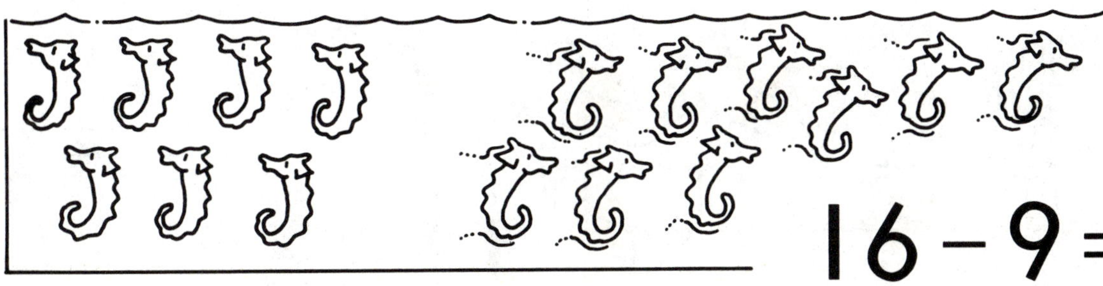

$16 - 9 =$ ____

$15 - 8 =$ ____

Use a sticker to answer.

What has many holes but still holds water?

Skill: To subtract numbers 0–9 from numbers through 18.

Nuts and Bolts

Subtract.
Write how many are left.

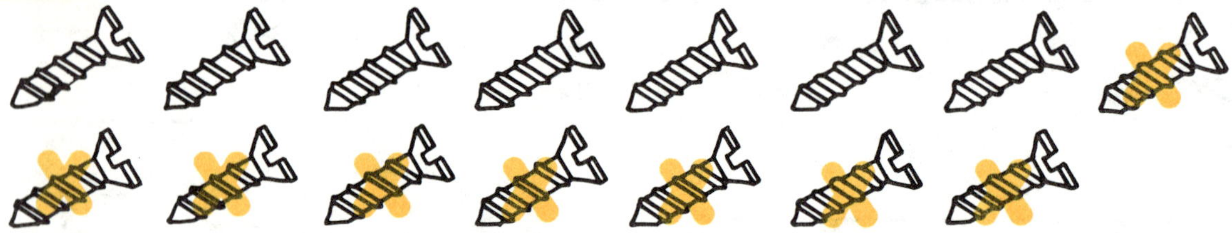

How many are left? $15 - 8 =$ 7

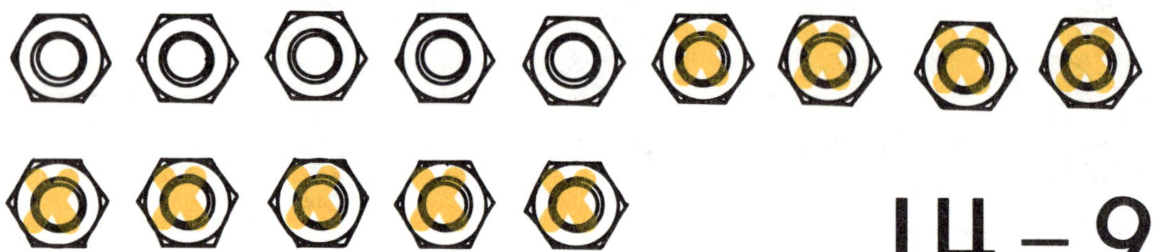

$14 - 9 =$ ____

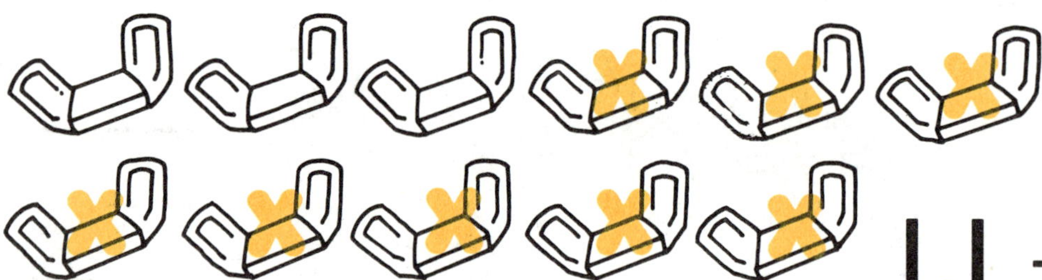

$11 - 8 =$ ____

$18 - 9 =$ ____

Use a sticker to answer.

What  **should you never**  **?**

 Skill: To subtract numbers 0–9 from numbers through 18.

Too Many Tools

X out.
Subtract. Write how many are left.

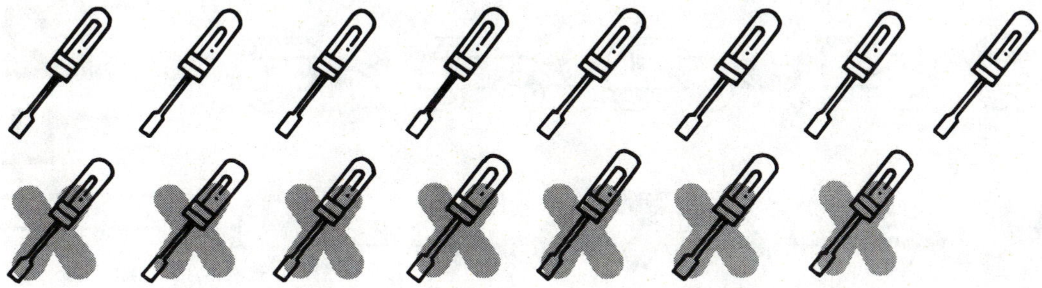

$$\begin{array}{r} 15 \\ -\ 7 \\ \hline 8 \end{array}$$

How many are left?

$$\begin{array}{r} 16 \\ -\ 8 \\ \hline \end{array}$$

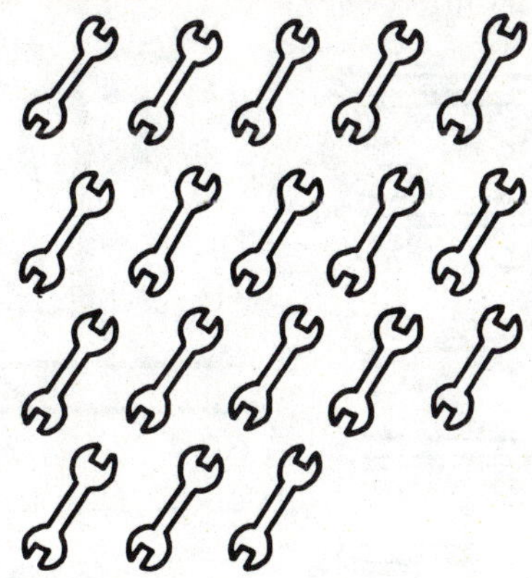

$$\begin{array}{r} 18 \\ -\ 9 \\ \hline \end{array}$$

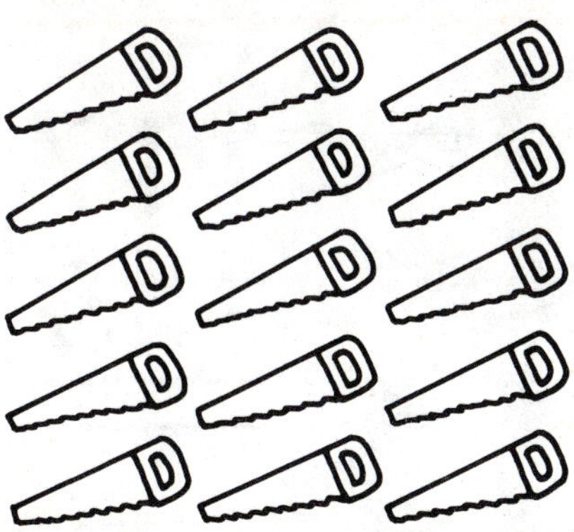

$$\begin{array}{r} 15 \\ -\ 9 \\ \hline \end{array}$$

$$\begin{array}{r} 16 \\ -\ 7 \\ \hline \end{array}$$

Find one sticker to answer all.

15 − 6 =
17 − 8 =
12 − 3 =

Zoom Along

X out.
Subtract. Write how many are left.

$$\begin{array}{r} 12 \\ -\ 4 \\ \hline 8 \end{array}$$

How many are left?

$$\begin{array}{r} 10 \\ -\ 4 \\ \hline \end{array}$$

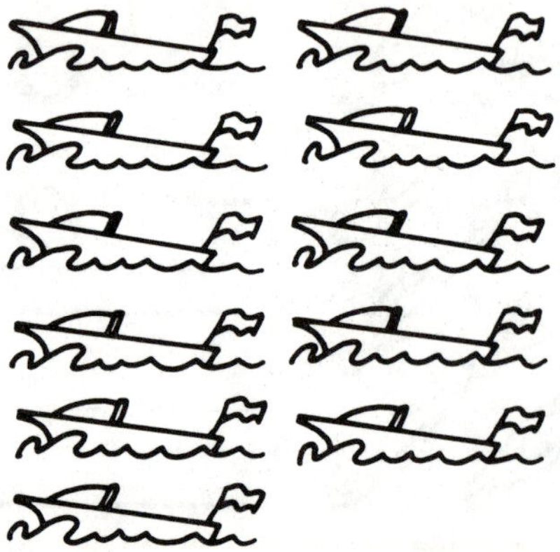

$$\begin{array}{r} 11 \\ -\ 7 \\ \hline \end{array}$$

$$\begin{array}{r} 16 \\ -\ 9 \\ \hline \end{array}$$

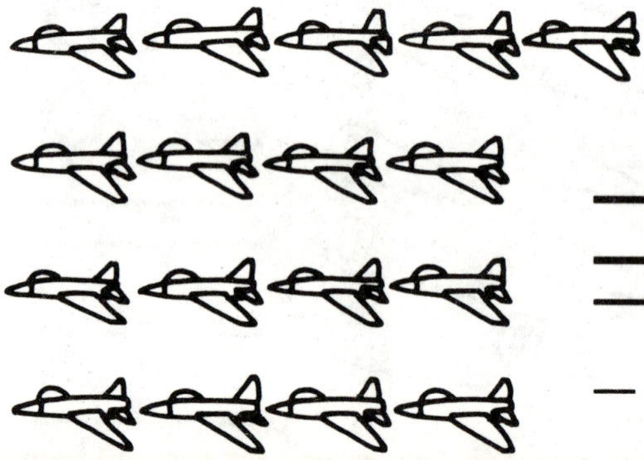

$$\begin{array}{r} 17 \\ -\ 9 \\ \hline \end{array}$$

Use a sticker to answer.

What kind of car goes without a motor?

To subtract numbers 0–9 from numbers through 18 in vertical form.

Lucky for them, a tailor came along. He sewed the beans together with his thread.

And that is why to this day all beans have seams.

The Straw, the Coals, and the Beans

Long ago and far away an old woman began to cook her supper.

The old woman had a bunch of 35 dry straws to light her fire. But 13 slipped out of her hand.

The poor straws began to burn, and soon broke in two. All the coals and all the straws tumbled into the rushing water.

The beans who had stayed on the riverbank watched it all happen. They laughed until they split their sides.

The straws threw themselves across the river. And the hot coals began to cross on the backs of the straws.

But when they got to the middle, the coals became so scared of the rushing water they could not move.

Soon the woman had 27 glowing coals to warm her pot. But 13 jumped out of her fire.

The woman put 19 fat beans in the pot. But 5 of these slipped away.

The straws, coals, and beans that got away went off to seek their fortunes. Along the way they came to a river with no bridge.

Teddy Bear Take Away

Subtract means **take away**.

Take away.
Write how many are left.

How many are left?

$$\begin{array}{r} 11 \\ -\ 2 \\ \hline 9 \end{array}$$

$$\begin{array}{r} 13 \\ -\ 5 \\ \hline \end{array}$$

$$\begin{array}{r} 15 \\ -\ 8 \\ \hline \end{array}$$

Use a sticker to answer.

Take 2 **.**
Then take 2 more.
How many are left by the door?

More Doors

Subtract to compare.

Match.
Subtract to compare. Write the difference.

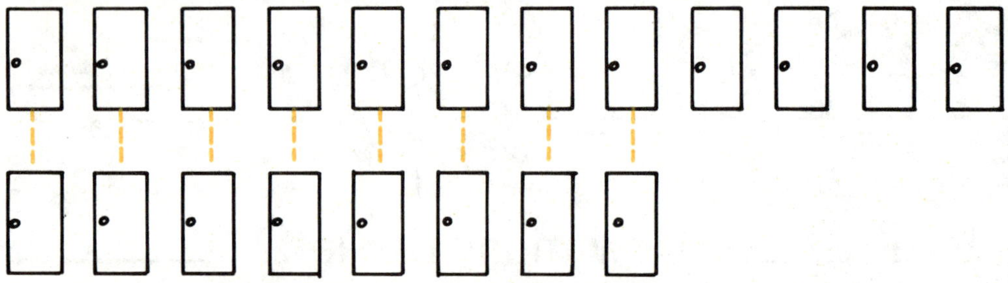

$$\begin{array}{r} 12 \\ -\ 8 \\ \hline 4 \end{array}$$

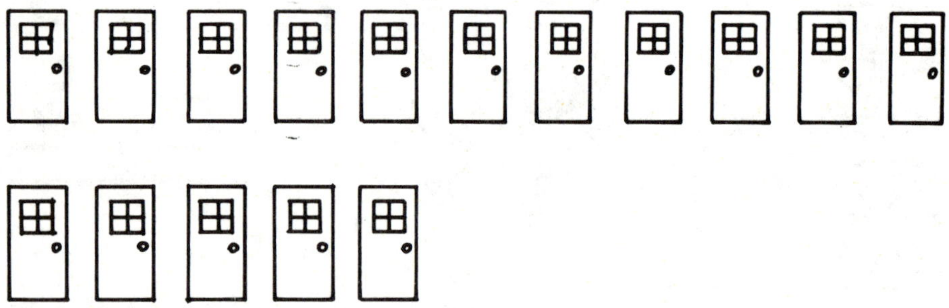

$$\begin{array}{r} 11 \\ -\ 5 \\ \hline \end{array}$$

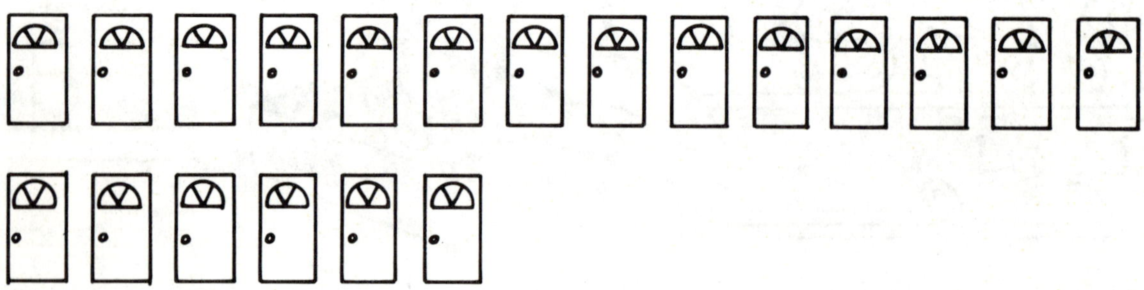

$$\begin{array}{r} 14 \\ -\ 6 \\ \hline \end{array}$$

Use a sticker to answer.

Where does this door belong?

Skill: To use subtraction to compare.

I Need Some More

Subtract to find how many more are needed.

Subtract. Write how many more are needed.

$$\begin{array}{r} 9 \\ -\ 5 \\ \hline 4 \end{array}$$

$$\begin{array}{r} 10 \\ -\ 7 \\ \hline \end{array}$$

$$\begin{array}{r} 10 \\ -\ 5 \\ \hline \end{array}$$

Use a sticker to answer.

 **has 8** **.**

He wants 10 **.**

How many more  **does he need?**

Skill: To use subtraction to find how many more are needed.

Time for Tens

Count by tens.
Find the sticker.

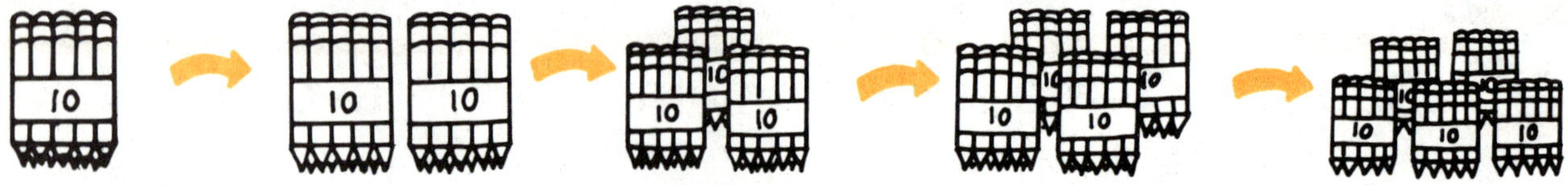

10 → 20 → 30 → 40 → 50

60 → 70 → 80 → 90 → ______

Count.
Write the number.

10, 20, ______, 40, ______,

60, ______, 80, ______, 100

Count back.
Find stickers for the missing numbers.

100, 90, ______, 70, ______,

50, ______, 30, 20, ______

Skill: To count forward and back by tens from 10 to 100.

Dimes and Pennies

A dime is 10¢. **A penny is 1¢.**

10¢ + 3¢ = 13¢

Count how much.
Write the numbers.

____¢ + ____¢ = ____¢

____¢ + ____¢ = ____¢

____¢ + ____¢ = ____¢

Use a sticker to answer.

Where do pigs keep their money?

Skill: To understand the concept of tens and ones using dimes and pennies.

Tens Take Away

X out tens.
Subtract. Write how many are left.

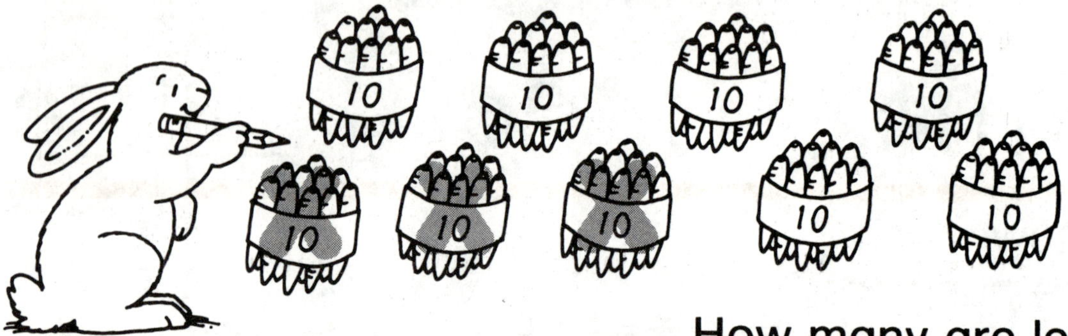

$$\begin{array}{r} 90 \\ -\ 30 \\ \hline 60 \end{array}$$

How many are left? 60

$$\begin{array}{r} 80 \\ -\ 40 \\ \hline \end{array}$$

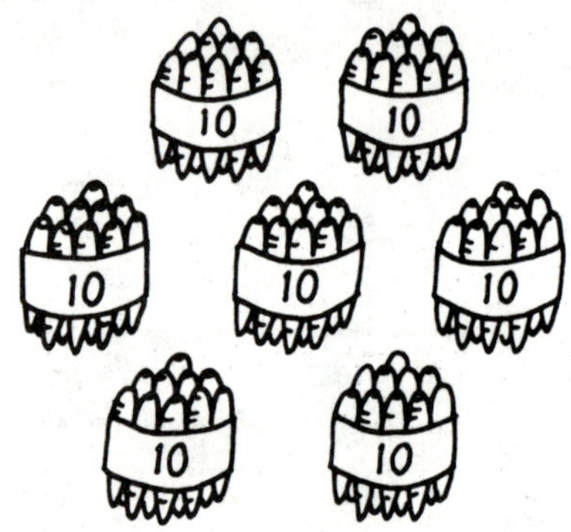

$$\begin{array}{r} 70 \\ -\ 50 \\ \hline \end{array}$$

Subtract.

$$\begin{array}{r} 60 \\ -\ 20 \\ \hline 40 \end{array} \qquad \begin{array}{r} 80 \\ -\ 60 \\ \hline \end{array} \qquad \begin{array}{r} 50 \\ -\ 40 \\ \hline \end{array} \qquad \begin{array}{r} 70 \\ -\ 20 \\ \hline \end{array} \qquad \begin{array}{r} 90 \\ -\ 60 \\ \hline \end{array}$$

Use a sticker to answer.

What tree is like a part of your body?

 Skill: To subtract multiples of ten.

Brush Off

Subtract ones.
Write how many are left.

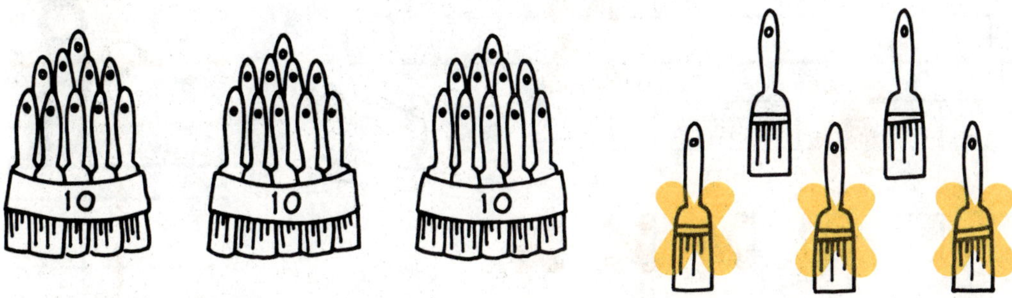

	tens	ones
	3	5
−		3
	3	2

	tens	ones
	2	6
−		4

	tens	ones
	4	8
−		2

87	54	63	99	75
− 6	− 1	− 3	− 4	− 2
81				

Use a sticker to answer.

What coat won't keep you warm? **a coat of** ______

Subtract a Lot

Subtract tens and ones.
Write how many are left.

tens	ones
3	8
− 2	3
1	5

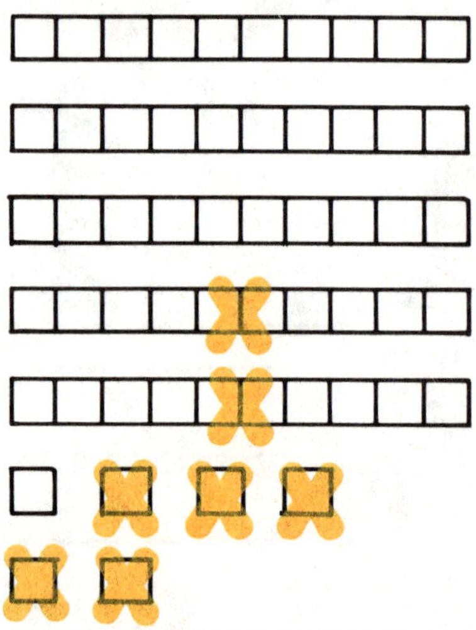

tens	ones
5	6
− 2	5

tens	ones
4	7
− 1	4

$$\begin{array}{r} 39 \\ -\ 28 \\ \hline 11 \end{array} \qquad \begin{array}{r} 65 \\ -\ 21 \\ \hline \end{array} \qquad \begin{array}{r} 98 \\ -\ 40 \\ \hline \end{array} \qquad \begin{array}{r} 73 \\ -\ 43 \\ \hline \end{array} \qquad \begin{array}{r} 84 \\ -\ 72 \\ \hline \end{array}$$

Use a sticker to answer.

What has teeth but no mouth?

Skill: To subtract ones and tens.

Know It All

Subtract.

10	12	18	16	12
− 7	− 6	− 9	− 7	− 8
3				

15	13	17	11	14
− 7	− 9	− 8	− 8	− 6

80	79	46	29
− 50	− 45	− 5	− 13

Use a sticker to complete.

I did it! I am a ______!

Skill: To review subtraction.

Subtraction Stories

Read the story.

Subtract.

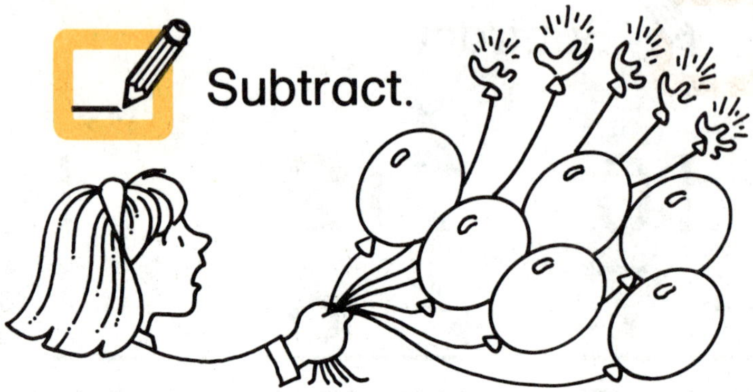

I had 11 

5  popped.

How many  are left? $11 - 5 = \_\_\_\_$

I saw 30 .

20  swam away.

How many  are left?

$$\begin{array}{r} 30 \\ -20 \\ \hline \end{array}$$

I have 6 .

I need 9 .

How many more  do I need?

$$\begin{array}{r} 9 \\ -\ 6 \\ \hline \end{array}$$

Use a sticker to answer.

What building has the most stories?

Skill: To solve subtraction story problems.

Money Math

X out.
Subtract.

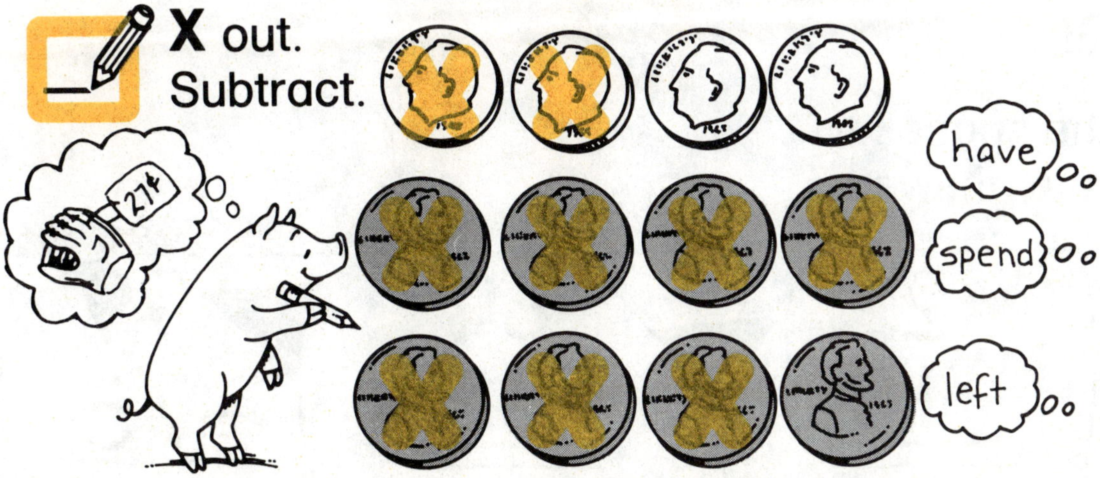

$$\begin{array}{r} 48¢ \\ -\ 27¢ \\ \hline 21¢ \end{array}$$

$$\begin{array}{r} 56¢ \\ -33¢ \\ \hline \_\_\_¢ \end{array}$$

$$\begin{array}{r} 69¢ \\ -42¢ \\ \hline \_\_\_¢ \end{array}$$

Subtract.

$$\begin{array}{r} 83¢ \\ -\ 50¢ \\ \hline 33¢ \end{array} \qquad \begin{array}{r} 27¢ \\ -15¢ \\ \hline \_\_\_¢ \end{array} \qquad \begin{array}{r} 95¢ \\ -21¢ \\ \hline \_\_\_¢ \end{array} \qquad \begin{array}{r} 74¢ \\ -63¢ \\ \hline \_\_\_¢ \end{array}$$

Use a sticker to answer.

I equal 2  **and 5** **.**

What am I?

Skill: To review subtraction of ones and tens using pennies and dimes.

What's Hiding?

Subtract.

Color the spaces to match the answers.

8 = blue 12 = green

24 = red 30 = yellow

 Skills: To review subtraction; to solve a visual puzzle.

Practice Test

I know my subtraction facts.

$10 - 5 = \underline{5}$

- ○ 6
- ● 5
- ○ 4

Subtract.

A. $18 - 9 =$ ____

- ○ 9
- ○ 8
- ○ 7

B. $13 - 7 =$ ____

- ○ 8
- ○ 7
- ○ 6

C. $15 - 8 =$ ____

- ○ 8
- ○ 7
- ○ 6

D. $12 - 9 =$ ____

- ○ 5
- ○ 4
- ○ 3

E. $9 - 9 =$ ____

- ○ 2
- ○ 1
- ○ 0

F. $14 - 6 =$ ____

- ○ 9
- ○ 8
- ○ 7

G. $11 - 7 =$ ____

- ○ 4
- ○ 3
- ○ 2

H. $16 - 8 =$ ____

- ○ 9
- ○ 8
- ○ 7

Practice Test

I can subtract.

$$\begin{array}{r} 40 \\ -20 \\ \hline 20 \end{array}$$

- ○ 30
- ● 20
- ○ 10

Subtract.

A. $\begin{array}{r} 60 \\ -20 \\ \hline \end{array}$

- ○ 60
- ○ 50
- ○ 40

B. $\begin{array}{r} 37 \\ -\ \ 5 \\ \hline \end{array}$

- ○ 32
- ○ 31
- ○ 30

C. $\begin{array}{r} 54 \\ -\ \ 3 \\ \hline \end{array}$

- ○ 52
- ○ 51
- ○ 50

D. $\begin{array}{r} 78 \\ -53 \\ \hline \end{array}$

- ○ 26
- ○ 25
- ○ 24

E. $\begin{array}{r} 89 \\ -43 \\ \hline \end{array}$

- ○ 47
- ○ 46
- ○ 45

F. $\begin{array}{r} 64 \\ -14 \\ \hline \end{array}$

- ○ 50
- ○ 40
- ○ 30

G. $\begin{array}{r} 32¢ \\ -11¢ \\ \hline \end{array}$

- ○ 22¢
- ○ 21¢
- ○ 20¢

H. $\begin{array}{r} 45¢ \\ -22¢ \\ \hline \end{array}$

- ○ 13¢
- ○ 23¢
- ○ 22¢

Skill: To test subtraction.

page 28

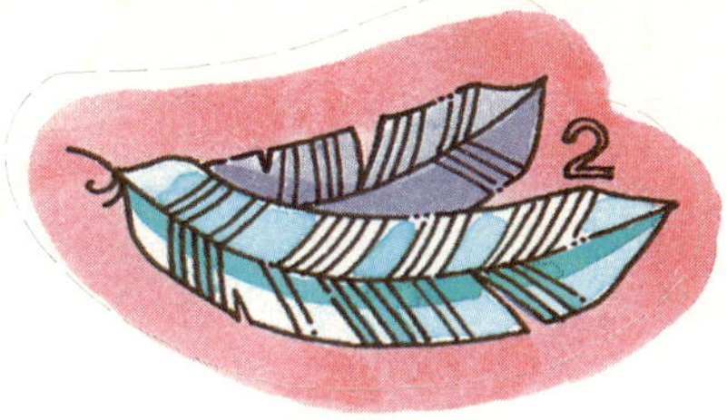

page 21

page 24

page 22

page 23

page 25

page 27

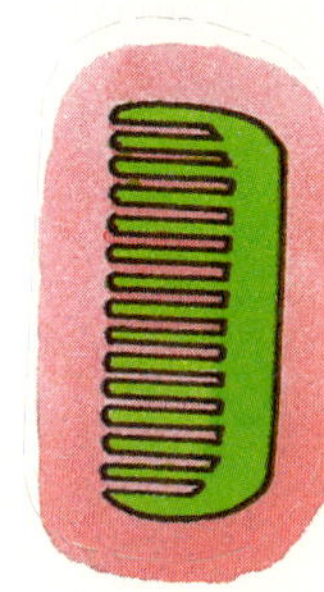

page 26

page 20

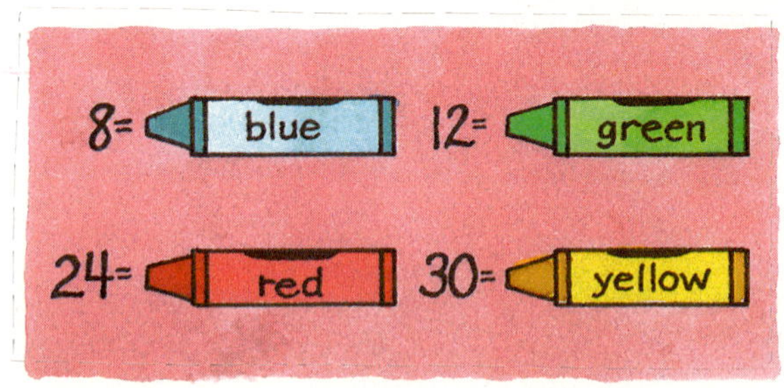

page 30

page 29

page 31

page 32

Extra reward!